BEFORE READING THE BOOK PLEASE READ THIS DISCLAIMER

The information presented in this book is educational in nature and is provided only as general information. As part of the information contained in this book, you understand you will be introduced to using a pendulum (an intuitive communication device) and provided information about various charts which can be used with a pendulum. By reading this book you understand the author and publisher do not know how you will personally respond to using a pendulum and whether your use of a pendulum and/or any of the charts will help you with a particular issue. You agree to assume and accept full responsibility for any and all risks associated with reading this book and using a pendulum and/or any of the charts contained in this book.

The information contained in this book is not intended to represent that the use of a pendulum and/or any of the charts are used to diagnose, treat, cure, or prevent any disease or psychological disorder. The use of a pendulum and/or any of the charts is not a substitute for medical or psychological treatment. Consequently, reading the book and using a pendulum and/or any of the charts for yourself does not replace health care from medical/psychological professionals. You agree to consult with your health care provider for any specific medical/psychological problems. In addition, you understand that any information contained in the book is not to be considered a recommendation that you stop seeing any of your health care professionals or using prescribed medication, if any, without consulting with your health care professional, even if after reading the book and using a pendulum and/or any of the charts it appears and indicates that such medication or therapy is unnecessary.

The author and publisher accept no responsibility or liability whatsoever for the use or misuse of the information contained in this book. The author strongly advises that you seek professional advice as appropriate before implementing any protocol or opinion expressed in this book, including using a pendulum and/or any of the charts, and before making any health decision.

By continuing to read this book, you knowingly, voluntarily, and intelligently assume these risks, including any adverse outcome that might result from using a pendulum and/or any of the charts, and agree to release, indemnify, hold harmless and defend the author and publisher, and their respective heirs, agents, consultants, and employees from and against any and all claims which you, or your heirs and/or representatives, may have for any loss, damage, or injury of any kind or nature arising out of or in connection with reading this book and using a pendulum and/or any of the charts. If any court of law rules that any part of this Disclaimer is invalid, the Disclaimer stands as if those parts were struck out.

BY CONTINUING TO READ THE BOOK YOU AGREE TO THIS DISCLAIMER

ISBN: 9781787233959

CONTENTS

INTRODUCTION

PREPARATION

MEDITATION

THE ART OF QUESTIONING

Q&A

AUTO-SUGGESTION

NEUTRALITY

CHOOSING A PENDULUM

SOME RECOMMENDATIONS

RADIESTHESIA or DEVELOPING WHOLISTIC AWARENESS

THE CONVENTION

THE HOW

THE STATE OF MIND

THE 4 STAGES

TRAINING

CHARTS, GRAPHS & TABLES

APPLICATIONS

CONFIDENTIALITY AND RESPECT OF BELIEF

PROTECTION

ABOUT THE AUTHOR

INTRODUCTION

It seems that a method to navigate between the dimensions of spirit and material reality would be most welcome. Whether it is a bridge to find a sense to the human experience of waking, dreaming and sleep; the physical and invisible spirit worlds; the cosmic and terrestrial. A bridge that restores communications which we have managed to lose, for whatever reasons. An access to truth and harmony.

For some, it could be astrology; for others, divining in some form or other. For me, it is radiesthesia.

Access to the truth, however trivial, however relative, can only be of benefit. The trick is determining how relative and that requires a firm understanding of what you are doing, of what is happening, and most importantly, if possible, a perfect balance in the role of the bridge.

In the final analysis, you are 'doing' nothing, you are merely open to the space where you are, you do not even have a choice in the matter, you are. You can allow the ego to chip in, but that is probably the extent of your true free will. So, do away with any notion of action, authorship, ownership, it can only get you into trouble and the false belief that you are some kind of cause/effect, or channel in whatever is going on, therefore distracting you from the task at hand.

A closer look at what might be going on

For ease of understanding, I will start with the currently accepted notion that it is the brain which deals with thinking. Further on in our studies, we will take a look at the traditional Chinese attitude towards organ function. It is of interest and quite revealing that a relatively modern western theory partially corroborates that it is the small intestine which plays a major part in this thinking process, rather than some zone of the brain.

The cerebral conscious element, the "mind" or intellect, is a powerful tool but with severe limitations because it functions in close collaboration with the emotions, those sensations acquired by means of individual experience and our education (parents, teachers, books, movies, music, etc.). This "mind" has nothing whatsoever to do with reality, it is a constant chimera of images, subject to personal interpretation, nevertheless sometimes shared with others, when we call it the world. There seems little point developing further the numerous ideas that abound around this subject. It has been done by others far more qualified and competent than I.

The real stuff of existence-consciousness (spirit) is universal, and that word contains a clue, uni, or the oneness that is the canvas on which all our feelings and perceptions can find expression. This is what I call the **all-conscious**. It is always there, nothing can affect it, everything has it and nothing can be without it.

By using a pendulum one is simply calling out to this all-conscious because that is where everything is deposited, including knowledge. They are not to be found in any of the so-called faculties of the conscious cerebral mind, or in the cells which are constantly dying and being replaced.

Call it what you will, it is not bothered by names. IT IS.

When working with a pendulum one needs to be mentally removed from the domain we normally operate in, we need - ideally - to be in the all-conscious zone, with no thing influencing our capacity of connection. For it is a question of developing another sense, so to speak. That connecting sense is always there, and always has been, we just forgot about it, that's all! Which is quite normal if the focus is transferred to the material, rather than the invisible force of spirit.

Answers to your questions are there but if the connection is clouded by the busy, external world-aware mind, the whole process is altered and you don't know, or worse, you know with impression-coloured spectacles. This remarkable capacity of thinking (generally wanting) can be turned to our advantage when we work with it, rather than being carried along in an uncontrolled stream. In either case, we have what goes by the name of auto-suggestion. If you feed your mind with thoughts of harmony and calm, they assume a form and space where you fit in. If you do not feed your mind in such a serene manner, it will be filled with emotional input which is generally of a misunderstood, chaotic and uncontrolled nature. So, be natural, think of nothing!

All perceptions of the senses apparently transit through the memory, adding to the psychic capital in the all-conscious, so forming the basis for emotional reaction.

What does that actually mean for us? From an early age we are told we must be active, we must stay busy, the idle mind breeds mischief, etc. The resulting noise which we religiously reproduce, in a knee-jerk reaction to this received wisdom, is a constant stimulus to the sense organs of perception, if not those of physical action. Only when we manage to calm – even for an instant – the mental flux, are we able to see it for what it is. Much ado about nothing. A constant kaleidoscope of sensory perception. This is where humans have the much vaunted capacity of free will. The ability to say, not for me the whirligigs and never-ending roundabouts. But you are a kill-joy, saying such a thing. That depends for whom. What does one want from life? A raising of the veil and a peek behind the scenes, or constant sensory stimulation, until one collapses from exhaustion in total unawareness or, even worse, ignorance?

However, when called up and perceived by the conscious mind these sensory perceptions are interpreted with the latter's limited power of reasoning, and its means of deduction are not equipped to find a solution, if deprived of the essence. The logical thing to do would be to refer to the all-conscious for help. But rest assured, humans are not rational, although that is what we are working on here.

This is why it is so important for the head and heart, the conscious and all-conscious "minds" to be in harmony. But if no one tells you, you can only stumble along, getting it right at times but rarely by design. It requires a very determined effort preceded by quiet conviction to understand that the all-conscious rules! We are a part of Nature, after all.

The reputed words of Siddhartha Gautama (the Buddha) come to mind: "Man does not have a soul." There is a partial truth in that, I say, the soul has a man. If only we can recognize the magic of spirit at work in one and all, we can approach Nature, instead of forcing Nature to comply to our whims.

Thanks to the evolution and development of science or knowledge of sense-recognized phenomena, rather than the function of knowing or gnosis, our connecting instinct has been steadily eroded and dominated. This faculty is, nevertheless, <u>always</u> accessible if the all-conscious has the necessary time to render an impression not imposed upon by the cerebral. We constantly call upon the all-conscious, albeit unawares, in everyday circumstances. After all, it is in a state of permanent waking, experiences no fatigue, nor error and consumes no mental energy. One only has to listen!

It is in the all-conscious that ideas are associated with the memory, playing a major role there for the individual. When you see something which reminds you of an event, something that happened in the past, it is because, independently of any deliberate effort, the all-conscious synchronizes that event to the memory it has in store. Without the all-conscious, the memory would be a jumble of unrelated souvenirs, only featuring in the conscious mind in an isolated manner. In the background it works quietly, secretly (to the conscious mind) relating and maturing the acquisitions of the memory and can, when called upon, offer them up to the intelligence, which is where the colouring book takes over, right/wrong, good/bad and we lose the unbiased simplicity of straightforward observation, familiar in meditation.

The radiesthesist is unable to live in a material world alone, everything has a potential reality or probability.

PREPARATION

The first obstacle, and surely the toughest one to overcome, is our ignorance. There is a vast difference in not knowing something and ignoring the same once you are aware of its existence. It is normal, even if uncomfortable, not to know something. Nature provides us with the perfect physiological system to work on that – namely our sensory function. We have many more than the five Aristotle dictated, for where he went to school in Egypt, the senses totalled 360, no known list of these exists, but no doubt included the Hindu "mind" complex, with its four aspects of thinking, intellect, memory and ego.

Would it not be perverse to neglect the harm caused by something and then to embrace that same thing. Human society has laws to prevent that type of behaviour, but no one will seriously suggest that humans are even able to govern their desires to the extent of staying out of harm's way. This is not a judgement, merely an observation which has a substantial impact on whether an individual is able to stay healthy.

That is what concerns us here, I think, what is to our maximum advantage. Not only to stay healthy, but happy and harmonious.

Our modern societal systems have been developed as a result of this deliberate ignorance and as a consequence of the financial rewards. So there is good reason to believe that an element of our suffering is due to our own moral inadequacy. If there is any hope to find a solution to that suffering we could do worse than to take a wholistic approach. Such a course must start with a serious revision of education, whether academic, parental, media-based, or hearsay; including all branches of science, history and religion.

Having said that, there is no point doubting unless you have a satisfactory means to find answers to the questions that arise.

That is the aim here: to provide a little background to this very simple method of finding an answer to a question.

MEDITATION

Much ink has flowed on this, and I'm not about to chop down a tree to add my grain of salt, which I cannot resist however, albeit a single grain!

According to the Vedantic tradition, meditation is a 24h/7 business. So different from what one might read or hear from the numerous sources available today. It is a question of witnessing or calm observation of what's going down in the neighbourhood – all modalities included (emotions, mentation, feelings, impressions, etc.). Practice of that sort, consolidated by contemplation and guidance of authoritative text and teacher, all help to create the conviction that there is a unique substratum to all our worlds and dimensions; the all-conscious being the zone where we, as radiesthesists, need to be.

Answers 'come' from that all-pervasive yet totally invisible place.

The way to access it, and not just accidentally or randomly, is to be familiar with it; and what better way to be at ease there than to hang out there?

That is where meditation can usefully get you.

Example:
Sitting quietly and comfortably in a natural setting, with the noises of Nature within ear reach, ask to hear your noise. The rustling of a leaf, the buzzing of an insect, the sound of a bird, the warmth of the air, whatever. Focus solely on what you are in connection with, allow yourself to be absorbed and to absorb the sensation for as long as it takes for you to be at one with it. Stay in that dimension for as long as you are at ease and nothing else distracts you.

The idea being to appreciate that there is no separation in actuality between the inside and the outside world. It is as if you are an extension/abbreviation in a continuum. Once one has experienced this sensation of oneness or unicity, it somehow becomes easier to connect with it on return to the 'real' world with the pendulum and your questions.

THE ART OF QUESTIONING

The question asked has to be prepared on a structured decision-tree concept. The question needs to be very precise. Any ambiguity in formulation risks meeting with failure in finding the correct answer. You must become a Sherlock to succeed!

This may sound complicated, but it isn't. Like everything else, you need to establish a method.
The way to limit potential mistakes is by careful thought before asking a question.

Perhaps an example will make things clearer.

You are not sure if the peanut butter you bought from the organic grocery store is good for your child, and you want to find out because that is what he wants to eat for breakfast. And the last thing you need this morning is to have to drive your unhappy offspring to the hospital because of a violent allergic reaction.

Obviously you can't be certain if the peanut butter is good or not, even if the grocer reassures you that the nuts are from a farm where they do not use chemicals. So, with pendulum in hand, you ask the question: "Is this peanut butter good for my child here?" Your logical, conscious mind/brain is in doubt and formulates this question, because an answer would solve the immediate problem. The question is directed to the all-conscious, or that part of the universe that is not cognizantly accessible to the 'five' human senses as we have physically developed them, but is presumably connected to everything in the universe via the shared principles of existence and consciousness.

The answer to what might be good or bad for this particular child's organism is known by the all-conscious because of its absolute space networking ability—to use geek-speak—and the answer is relayed to the nerves, which then cause the muscles to react in a certain way.

The important aspect here is the question: "Is this peanut butter good for my child here?". Let us analyse this from the angle of semantics. What do we mean by the word 'good'? Obviously, in this example, it is the context that determines the sense. That is not always the case in the questions one asks, so it is vital that the sense is clear in your cerebral mind for the message to be received by the all-conscious.

In the final analysis, the word has a number of possible senses, which could easily change the answer:
a/ Good in that the peanut butter is not going to cause harm (the idea of hospital is at the back of your mind).
b/ Good to taste.
c/ Good with regard to nutrition. In which case the answer could well be negative.

d/ Good to keep the brat quiet for a while!
e/ Good with regard to the intrinsic qualities of the foodstuff.

All these possible contingencies must be considered, there are perhaps more!

Let's have a broader look at this aspect.

The course of our life is very dependent on our priorities. Very often in the bustle and confusion of the moment, we are not one-pointed enough in the phrasing of our intentions. To avoid the frustration and waste of time that will inevitably accompany vague questioning, getting to grips right from the start of this adventure would be an excellent initiative. I dare say you will find this same discipline applies to other domains where you are called upon to be precise in the formulation of a question; I think especially of horary astrology. Some examples will speak more eloquently.

Think of a situation in your life where there is a doubt as to the benefit of your possible action, or lack of it.

- A – Should I have another cup of coffee?
- B – Is there any danger in having another coffee?
- C – Can I afford another cup of coffee?
- D – Will another cup of coffee be good for me?

In question A, there is no reference to the benefit or otherwise, simply a question of quantity.
In question B, there is implied reference to potential harm being caused.
In question C, there is vagueness; are we talking about money, health, or time?
In question D, again a matter of quantity from a vague angle of "good".

Question B would be the least vague, settling the doubt – established in your original proposition to resolve the situation – as well as providing a conclusive answer to possible benefit, or not.

Question and Answer: Q & A.

In which the operator must learn to use faculties of intellect and intuition, applying either at will and never confusing them - the intellect for the formulation of questions and the evaluation of answers, and the intuition, using the radiesthetic faculty, to obtain the truth. Q & A is eminently the instrument of scientific radiesthetic research.

This I regard as the most important use of the radiesthetic faculty as it provides a bridge between two worlds - the sensible and the supersensible.

The elements of seeking and finding are of course inherent in all radiesthetic and dowsing work, but it is only in Q & A that they become a deliberate technique, and there is conscious "asking".

From a lecture given at the Congress of the British Society of Dowsers held at Malvern, 5th May, 1972 by Dr. Aubrey Westlake.

"That the whole process of the art of Q & A is a vastly larger subject than a means of obtaining specific information in any one specialized field. I see it as a means of integrating the personality and of learning how to construct a bridge between the conscious and unconscious worlds in relation to life as a whole. In short, Q & A can be used as a process of self-development."

Mrs. Jane Wilcox in her closing lecture entitled "Question and Answer", with the intriguing sub-title of "A Bridge Between Two Worlds", at a Conference of the British Dowsing Association.

The Q&A aspect is generally the Achilles heel in radiesthesia. I cannot stress enough the importance of the just formulation of a question, after careful consideration as to the wording. Try your very best to avoid any possibility for a gap or opening that allows misunderstanding, ignorance, misconception to creep in to the equation.

If it does, as risks happening at some stage, realize that in all probability, it is due to faulty questioning.

AUTO-SUGGESTION

This is a vital notion in radiesthesia, it is both one of the most important components, yet one of the major dangers. Psychologically, suggestion (or formulated desire) is the introduction in the mind of an idea that transforms into an action within a period of time. The effects of suggestion are infinite as it is through their intermediary that we act on ideas, sensations and actions.

Auto-suggestion is a suggestion made to oneself, you transform an idea into an act.

The best time to practice auto-suggestion is when the idea is totally assimilated in the mind as being the best solution, formulated with the full consent of your will and in the absence of any doubt whatsoever as to its possibilities. One must have absolute confidence in auto-suggestion, but to accomplish that, the terms used must be condensed into a simple, carefully worded formula. If the formula is not clear, imprecise or ambiguous with regard to intent or desire, the potential is reduced proportionally. The best material conditions for concentration of thought and determination should be applied, namely away from the excitement of the external world, probably in silence. Whatever is good for concentration, preferably no physical discomfort, pains, noises, distractions or extreme temperatures. Probably the best time for this is prior to going to sleep and in order to reinforce the process, first thing in the morning. The advantage of such timing is that the all-conscious will review it during the night, when the conscious is absent, and in all probability put you right if there is anything amiss.

Efficient auto-suggestion requires:
1/ A single idea
2/ Just one principal auto-suggestion at a time
3/ Attentive repetition of the formula.

If repetition becomes tiring, reformulate the auto-suggestion into secondary auto-suggestions. Based on one's introspection in meditation and contemplation, determine the causes of your intellectual failings, the influences that distract and the relationships between these influences and failings.

The topic is now becoming very subjective, and it is not for me to stipulate what or how. I can only relate my own experience.

When asking a question with the pendulum, an impression often arose on the screen of my mind. Those impressions, *samskara* in Sanskrit, are the result of our education, the mental input of years of experience. Such a process is essential for the professional activity of being a human; that's how you know not to walk out into the street without looking left and right.

But that does not apply here, not as a radiesthesist. You MUST be in a unique state of mind that is the sole guarantee for the correct answer.

To ensure that you have a clean slate with no "I think I know the answer to this one" requires a constant purification process. It is a subtle process. It is best accomplished on the backdrop of the Hindu formula, *neti, neti,* (not this, not this) the practice of gradual elimination allowing the all-embracing harmony of neutrality.

NEUTRALITY

Neutrality is the secret in avoiding the influence of external suggestion and autosuggestion. If one can find suitable formula, and simply say: "I am totally neutral," "I know nothing," "I am looking for....," or "I would like to find...." you might well find that the preformed impressions take less space on the horizon.

This is why it is important not to linger too long on a question. So many ideas can jostle in the mind space just when they shouldn't!

The gravest error one can make, because it will invariably alter the answer, is to have the slightest belief of knowing what the answer is.

The above procedures are in large part adapted from Antoine Luzy's works, especially *La Radiesthesie Moderne*, 1943.

Tip: One of the most useful ways I have found to overcome this very normal tendency of 'thinking I know the answer' is to hold something in my left hand – the pendulum being in the right hand. A stone for example, and at the moment of repeating the question, I transfer my focus to the stone and leave the all-conscious in peace to sort things out as only it can!

Suggested exercise:

When seated (with eyes closed) in a calm, isolated place, think of a material object – not a person to begin with, a picture for example. You visualize the picture from the description or from what you have seen. Your whole undivided attention is concentrated on this alone, to the exclusion of all other ideas. Maintain normal breathing. When the outside world has completely disappeared, leaving only the idea and image of the picture in your mind, all things foreign to the search – noises, smells, etc. affecting the senses being absent, you have achieved the desired state. Maintain that state for a period specified beforehand, perhaps five minutes. This is a good way to achieve mental neutrality, required in the use of the pendulum. Depending on the individual, train every day until the state comes easily and quickly - for two weeks to three months.

CHOOSING A PENDULUM

Some home truths about pendulums right from the start are in order.

One can basically use anything to obtain an answer to a question one asks. A piece of string with a weight attached, a ring or simple metal nut; a blade of grass if it can move with sufficient flexibility, divining rods, and so on and on. The medium does NOT contain the answer, so should NOT be considered as an oracle! Unfortunately, many people invest a property into something which is no more than a vulgar medium.

How often does one hear that your pendulum is personal and you should not let anyone touch it? Do such protagonists also say you should not breathe the same oxygen as them, you might influence their thought! There lies the problem, there is no thinking in these matters which can only be seen as an attempt to control something which is not ours to control. Ignorance and her brother Mystification are alive and well!

This does not mean that a crystal, stone or precious metal pendulum does not have its individual properties, **BUT A DIVINING CAPACITY IS NOT ONE OF THEM**. The capacity to answer a question could well be said to belong to the supreme intelligence which governs our worlds, and although that intelligence is all-pervasive – even in the pendulum, it is the human thought, and the projected motion that thought generates which is necessary before answers are forthcoming.

There is a capacity, belonging to those few, very determined individuals, to invest an object with a specific attribute but that is a practice in the domain of "magic", which is not the focus here although often confused in the minds of those who tend to amalgamate what they find hard to explain.

So, the pendulum is a simple means to accomplish an end. Nothing more.

Having said that, there <u>are</u> pendulums of a very specific and precise design which due to their form or composition do have very unique capacities. The first that springs to mind is the Universal Pendulum created by André de Belizal and Leon Chaumery, or the Karnak pendulum – origin unknown, which has the unique property of creating what is known as the negative green frequency in the electric phase. That is the pendulum that has my preference for "exorcism" or spirit release. It is as if a soul can "ride" to the light and exit the "world of incarnate beings" on such a frequency. These pendulums are extremely useful, need to be very carefully designed, made and even more carefully handled. The Karnak must be kept unassembled because if not, it generates what de Belizal called (and patented in 1936) the "negative green" frequency. This frequency has an extremely short wavelength and is potentially the cause of the dehydration capacity which it is claimed to produce.

These French pioneers stated that this frequency is produced from the centre of the south wall of Kheops' pyramid, as well as in the King's chamber of the same structure. It has a wavelength smaller than an X-ray and is immensely useful in the magnetic phase, as it gives life, but quite dangerous in the electric phase, where it takes life. Incidentally, questioning with the pendulum might well be in order if you are considering building and/or using a pyramid. For some forty years I used a scale-model Kheops' pyramid to sharpen my razor blades; I had not thought of asking if there was any harm in doing so.

Once again, the pendulum reveals some useful information. Ask if your scale pyramid is beneficial or not. The answers I received, when asking if this harmful frequency generated by the large scale versions in Egypt, Kosovo, Bermuda and elsewhere, is dangerous for humans, were intriguing. Apparently it is, but when the structure is incorporated into the earth, the earth manages to absorb the force, but not so for the mobile versions.

The French radiesthesists, especially André de Belizal and Chaumery demonstrated that geometric form and shape produce remarkable results with regard to resultant frequency. Given that their research was generally based on the study of hemispheres, circles, spirals and the natural properties coming from these, there is every reason to believe that more ancient peoples were familiar with those properties, and that they used them for a variety of purposes.

SOME RECOMMENDATIONS

Practice is the key to becoming proficient and confident and you will find that the art of questioning is as important as the ability to move into the right state of mind while waiting for the answer to arrive from the all-conscious.

There are a number of "rules" of a practical nature which I have found best to observe to avoid making mistakes – which can still happen.

Physical vitality and moral rectitude are essential, so if you are feeling tired, depressed or upset it is best not to ask anything but the simplest of questions. By simple, I mean "Is it alright for me to eat this yogurt?"

The emotional aspect is extremely pertinent. If you are emotionally involved with someone, you risk finding erroneous answers when you ask about your partner/child/parent. Our thought process is clearly clouded by such issues, and however hard one tries to step away from them, the turbulence is nigh impossible to clear. Best to avoid working on your spouse, partner, etc. If you do need to do so, you might well consider using cards, one with 'yes' and one with 'no' written on it. Place the cards in two separate envelopes, shuffle the envelopes and lay them on the table. Ask your questions to the envelopes, noting the answers – left or right. When finished, open the envelopes and you will have the next best thing to a double-blind test!

Another aspect of the emotional component is whether you can handle the answer. For example, if you are very attached to a person and want to spend your life with them, asking if they are the right person for you, is fraught with danger. The same applies when you ask such a question for others. Be sure of your ability to deal with the answer, the responsibility can be great.

Asking into the future is a delicate matter, firstly there is too much that can change between times, secondly it may not be judicious for us to know, and even if we are really good at this business, that does not mean we can replace the supreme intelligence. If the questions are motivated by any sense of gain, you do so at your risk.

Having said that, if in the process of working with someone there arises the question of, for example, the sale or purchase of their house, and it is key to their future, it seems quite justified to ask at what moment should the transaction take place, and for what amount.

Many of the French authors on the subject of radiesthesia mention that it is best not to work on stormy days. There is of course magnetic and electromagnetic interference in such weather and I believe that is the origin of such concern, but if you are working from the standpoint of this whole operation being a purely conscious-to-all-conscious connection, there is little room for the weather, otherwise you would need to check for solar flares and coronal mass ejections, etc. You can always ask!

A word about 'healing'. A great many people these days seem to be involved in pendulum-healing, or the healing arts. Once again, I don't expect you to share my opinion, but I DO know that if you think you can heal someone or something, your ego risks being boosted, big time! My firm conviction, as a very ignorant but well-meaning cipher in this vast expanse of space, was so elegantly expressed by Amboise Paré, a French physician in the 16th century: "I patch things up, but God heals". You might agree, it would be a shame to put him out of a job.

———————————————————

RADIESTHESIA or DEVELOPING WHOLISTIC AWARENESS

Radiesthesia is much more than a method for finding answers to questions.

An image springs to mind perhaps - the guy with the hazel twig looking for water; the soldier with the rods looking for tunnels; or yet again, a person sitting at their desk, pendulum in hand, bent over a map searching for mineral deposits in some distant land. Those are not just possibilities. They are actual. Much like the astrologer working in a bank, deciding the right moment to invest, in what and how much.

No matter what <u>your</u> objective, if you are seriously considering adding this discipline to your toolbox, and obtaining the right answers to your questions, it would be best that, right from the start, you approach the process as a discipline. There are no two ways about it, if you hope to achieve the confidence that makes the difference between a good and indifferent radiesthesist, you need not only an awful lot of practice but a constant awareness of communion – communion with the essence. This is a divining art. We know, in all humility, that we are in a state of total ignorance, which must presumably be the case if we do not 'know' if the apple is nutritionally good for us, or just good taste-wise.

You are working with what <u>really</u> exists, not with the mental constructs which we often take for our reality.

It is a complex process, when viewed from the viewpoint of mental implication; it is even more complex when viewed from the physical standpoint. That is perfectly normal, because we do not know very much about anything, which is perfectly normal, we are part of it.

Most people who use the pendulum share the belief that everything that exists has its specific frequency or vibration. Now if you relate to that frequency, and why not, you are in a direct link with that object, and by definition no longer "connected" to whatever else is in the vicinity. This approach is known as the physicist school of radiesthesia, very closely related to modern science, where everything is dissociated and inspected in its own light, so to speak.

The mentalist school, as developed by the French in particular, works on a much more psychological principle, whereby the intellectual faculty of the mind asks a question because it does not know the answer, and thanks to the same neuro-muscular response, a reply comes from some unknown entity, but probably the subconscious.

The antagonism between the two is easy to understand, and is perfectly mutual.

If, on the contrary, you consider yourself, as I do, to be a part of, rather than apart from, everything that exists, the process becomes remarkably different. As of the time that you recognize that you don't know, you are potentially open to whatever is actually there. But if you focus on a specific frequency, you again create a subject-object relationship. That is not where one needs to be. It seems reasonable to say that the best possible position to perceive is in a state of vacuity, with nothing imposed on the mind space environment. There is a greater chance of connecting with that environment (and all to be found therein) when you feel you are a component rather than some sort of controlling influential factor.

Neither the mentalist nor physicist schools - as defined by the French in the early 20th century - fit this open mindset. The two are mutually exclusive, and both fall short of the total reception of the required perceiving. What is worse, to my mind, it does not correspond to a wholistic approach where inclusion is the key. Especially inclusion of one's thought process or mind, as one of many similar components that make up the human segment of this dimension. Which although highly important to us, is of little consequence to the flow of events from a cosmic viewpoint!

At some stage, and it will probably be early on, once you have acquired a certain ease in using a pendulum, you will ask " *What's going on here?*" I make no claims to knowing any more than the next person. In fact, on the contrary, I am such an ignorant fellow that I have to use a pendulum to discover answers to the simplest of questions.

The passage of time has taught me one thing.

Things change. Constantly.

But we do have this remarkable means of discovering the truth of the moment. Would it not be foolhardy not to access a possible answer if it is so readily available?

An explanation as to how a neuro-muscular response to a mentally formulated question can occur would be welcome. Or, at least a theory as to its apparent working.

There is one thing common to life - all life. It is consistently subject to natural forces. No matter how hard one tries to work with or against those energies, they resist. They have a life force of their own, in the same way that we do. One might even say they have an intelligence. Indeed, they do, but that is not what science would have us believe, for if anyone who has studied and worked with water will tell you, that is the only honest conclusion one can reach. When one stops being a person, a scientist, a dowser, a compartmentalized individual, there arises a chance - I repeat, a chance - that one is. "Is" as in "I am". In a materialistic regime, for the rest of our time here on earth, that angle is reinforced but rarely examined. In a more spiritual, wholistic regime, that is the angle to be examined and pared down to size.

The natural conclusion is a complete unity, a perfect totality with everything in its place, in constant but impeccable flux. Thank goodness, we do not see this as it is, it would be far too frightening for our fragile constituencies.

The whole point, however, is that there is a component which we perceive and feel, and another which we merely feel. The conscious and the (rather rudely termed) unconscious.

Radiesthesia is the bridge between these two: the phenomenally aware, conscious component, which might be equated with the cerebral mind, whereas the real brain is probably the entire human metabolism; and the much broader, because englobing all components of our physical reality, all-conscious (as I prefer to call it).

Of course, there is no differentiation, therefore no bridge, but our human ego separates you from me, and we spend a life-time trying to establish that bridge between two continents that were never apart in the first place!

If one can accept such a theory, there is a substantial advantage in that the phenomenon of dowsing does not have to depend on some elaborate explanation of physics (based on differentiation) or magic, which although it always will be, you no longer have to rely on some undefined, unknown – therefore mysterious, factor.

It is this mysterious side which gets any 'unscientific' phenomenon a bad name. If you cannot verify something, it is very easy to make a claim and attribute it to some peculiar property of which you are, of course, the sole guardian.

This approach to radiesthesia, based on communication between the conscious and the all-conscious, removes all need for any mystification, and at the same time puts mystification exactly where it should be. In the realm of egocentricity and real pseudo-science!

An anecdote if I may, *Kuan shih yin*, 'he/she who perceives the sounds of the world', Kuan Yin (the goddess of compassion) for the Chinese, or Avalokiteshvara, the Buddhist bodhisattva taken in the pantheon by the early monks to China some two thousand years ago, was the first sex-change in religion. For the martial Chinese, compassion is a female trait so the bodhisattva must be a woman rather than a man, and so it was. An interesting notion relevant to radiesthesia, which is the perception of waves/frequencies, a term invented by Abbé Boulé in 1920 or so. That is what we are trying to do: perceive things so as to be in harmony with what we find.

There might well be something of all this to be found in ancient cave drawings. Whether from Australia, China, France, Patagonia, South Africa or Spain, the paintings often portray a connection established between the subject - humans in the case of cave art (human hand-prints, hunters with bows, spears and assorted weapons), and the object – animals, or more specifically – lunch.

As with so many aspects of life, some people are gifted. Probably in those days – forty thousand years or more ago, the same held true. Some men or women were better able to locate the group of animals to be found – or avoided. Those individuals were no doubt encouraged to hone their capacity and were surely very useful members of society. Is that perhaps how they consolidated their skill, on the walls of their homes?

The connection enabling a link between an unseen source of life and our frequently urgent need to access the information as to the whereabouts of that energy source, is as real today as it was in the past. The means of tapping into it have changed, and in the process, the natural ability of "primitive" humans has declined. As often happens when a sense is not used, it atrophies and access is obscured.

That does not mean the information is no longer accessible. On the contrary, I believe, it is and always will be. But this 'sense' needs to be acknowledged, nurtured and developed.

Though it may sound trite to say so, we humans are equipped with a logical reasoning capacity. That is why we know when we don't know something. We have an ability to learn from our mistakes – well, some of us anyway. It is my firm conviction that we, as humans, in this modern age, do not question sufficiently. Not only do we not question whether something is right or wrong, but we do not doubt what an authoritative voice tells us. That is dangerous and tragic. It is as if we have lost the ability to discriminate what is good for us, or not.

In a manner of speaking, that is what has happened. We no longer have the know-how to obtain an honest answer to a simple question. The brain rules and we are led by the nose along the path of our existence that can only lead to our undoing if we fail to learn that it is the heart, our intuitive connection, which precedes. Would it not be more sensible to follow that path in the knowledge that it is the right one for us – individually or collectively – thanks to our capacity to discern?

That is the faculty we are now going to look at.

The pendulum is surely the most flexible of all the tools that can be used to find answers to questions. It is *par excellence* the "barefoot" way to discover if something (food, drink, person, medication, situation) is beneficial or not for you or others. Because of its flexibility and simplicity of use, the pendulum rather than other methods, such as dowsing rods, the hazel twig or equivalent, or again the Bi-digital O-ring, friction, arm muscle, sway, toe-touching tests, although equally valid, shall not be considered here.

There is very little that cannot be determined using a pendulum, and not a single domain in our physical life where it cannot be applied, though becoming confident in its use requires a lot of work; involving not just practice, but understanding, if one is to acquire the desired intuition. However, almost everyone can become proficient in its use.

There are a number of human qualities that would be well to acquire beforehand. Firstly, because the answers can only come one at a time, and only in the form of yes or no. Be kind on yourself, take your time! The overall picture will be completed with patient questioning but the process remains binary at all stages.

The very first step in radiesthesia is to decide what is yes, and what is no.

THE CONVENTION

The convention is your personal code, the method in the movement. Before starting to work with a pendulum or divining rods you need to determine what is "yes" and what is "no". There is only one strong recommendation here whatever you choose, once you have decided, <u>do not change</u>.

It could be a clockwise and anti-clockwise rotation of the pendulum? A lateral, north-south, sideways, east-west swing? From personal experience and use, often in strange places – such as in the dark when you cannot see, only feel, the movement, or in bad weather when the movement is influenced by the wind, the first option wins, but your call!

The same rule goes for all methods of dowsing - although with the rods, there can only be a crossing of the two rods or a stationary position, once you have decided what is yes, do not change.

There are some dowsers who opt for a 'maybe'. This would imply an incomplete or ambiguous question. If you introduce your indecision into the process, there will be a natural tendency for it to appear in future questioning. I am of the opinion that should be eliminated once and for all. I cannot recommend strongly enough that 'maybe' should be avoided. It is not a game of chance.

Having decided on the expression of yes and no, the next step to be considered is physical posture.

THE HOW

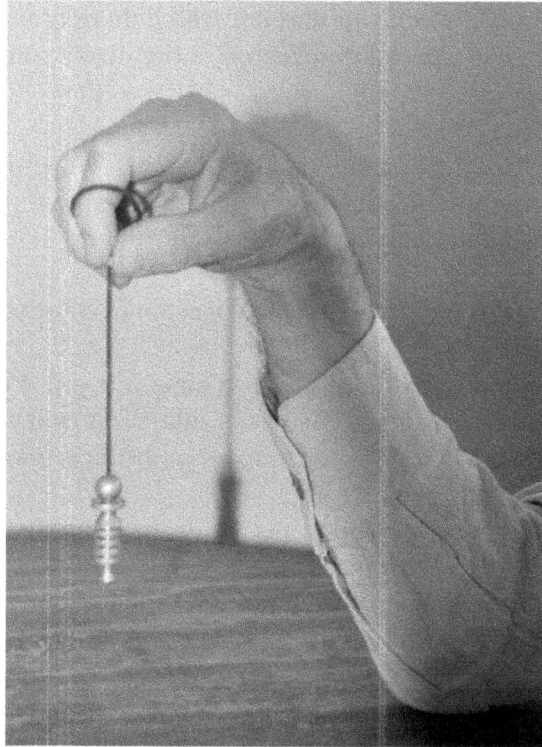

Sit comfortably at a table or desk, with the pendulum in hand - always try to use the same hand, the individual's bio-magnetic charge plays a part in the equation, so best to keep it simple and use only one hand. Rest your elbow on a firm surface, allowing your wrist to be completely relaxed – no tension – so that the pendulum is hanging down held between the forefinger and thumb. Tension on the string can interfere with the pendulum movement, so keep contact between the string and fingers to a minimum. Some pendulums come with a bead on the string, by holding the bead and not the string, you free the string of contact.

Be at ease, mindless, because response time may vary, especially to begin with. There is a similarity between meditation and using a pendulum in several areas. One of which is keeping your back straight, it facilitates the nervous flow, so it is best to sit upright. The least distractions, whether visual or audible, the better. To begin with, always assume the same position with your body until you feel confident that an answer comes regardless of your physical posture. The state of empty-mind is a huge benefit.

It is exciting the first time you feel the pendulum moving and there is a good chance that you will always feel a sense of awe. Stay humble and grateful.

THE STATE OF MIND

The mental process is simple, when you start a research there is an intention, a well-defined purpose known as the mental designation of the object sought. What exactly do <u>you</u> want?

The mental designation fixes the purpose of the research, focusses it and accompanies the desire to find what you want, this desire is formulated as an expression of an auto-suggestive nature which triggers the mental activity and solicits the all-conscious.

Let's take an example: You want to know if the yogurt in front of you is good for you?

Don't allow the TV hype and your friend's taste to influence you. You are not sure if there is MSG in there? You have an allergy to goat's milk? You don't like certain brands of yogurt? Many reasons may exist for you wanting to know if it is good for you.

But, you don't <u>know</u> if it is good for you or not. Now, in your current state of health. Is it the right thing for you to eat? Eliminate all the details for or not eating it. KEEP IT SIMPLE SUNSHINE (the US Navy invented the KISS acronym and were never known for their finesse, so Sunshine it will be!).

Your intention must be perfectly simple – to know if the yogurt is the right thing for you to eat – so keep the expression of the question simple and only allow that sole idea to exist in your mind. Again, the practice of meditation is useful as it enables this kind of focus more easily.

Focussing on the yogurt, looking at the package or contents in your bowl, ask the question, "Is this yogurt good for me to eat now?"

Go vacant! No thought, especially concerning the result. It will come, probably quite quickly but give it space. The pendulum will start its movement and you are in the know.

You have three ways of working with your yogurt! I would recommend, to start with, playing with the first two. Once you gain confidence, try the third, but always feel at ease with your method.

The methods are as follows:

With your pendulum hand resting on the surface and the other hand in relation with the yogurt:

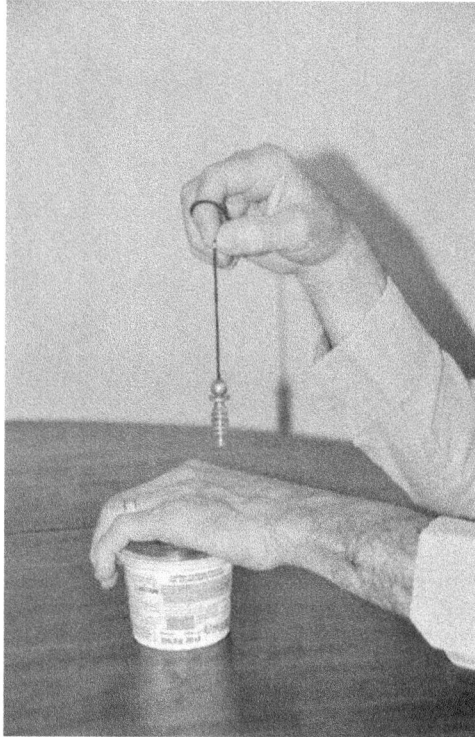

1/ Place your other hand over the container and ask the question. The "yes" or "no" answer will come in the way you have practised.

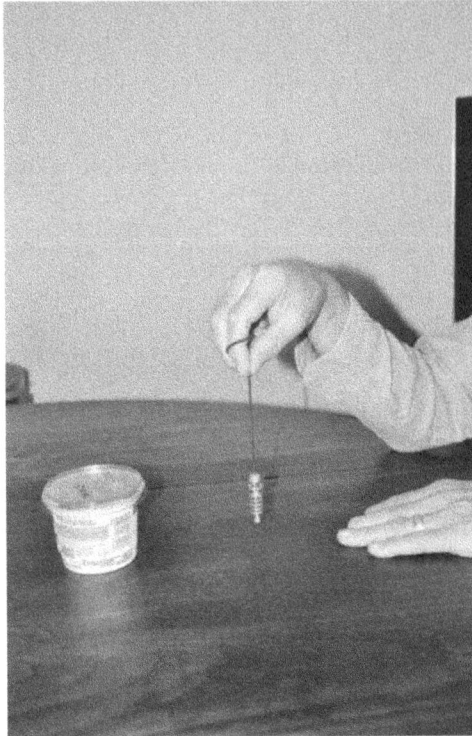

2/ Place your other hand on the same surface as the yogurt with about 20 cm between them. Ask the question with your pendulum-hand between the two. The pendulum will swing either in a line connecting your hand and the yogurt, in which case it is good. Or, it will swing sideways, breaking the connection, it is not the right thing for you today!

3/ Simply look at the yogurt or have it fixed in your mind's eye, ask the question. The answer will come in your chosen manner.

THE 4 STAGES

To resume methodically what you need to do, here are the four stages of the mental process. Here is an initial summary:

1. Mental designation of the object searched for, mentally envisaging and then orienting the mind towards it. One needs to define quite clearly what it is one is after, in as simple and precise terms as possible.

2. Formulation in so many words of exactly what you have just defined as the object of the research. It is the expression of your aim.

3. The passive state of neutrality. Having formulated exactly the objective to be achieved, the operator's all-conscious gently shifts into alert mode, a state of calm.

4. Stating the question.

1. Mental designation
The mental designation of the object searched for, directs and orients the operator's awareness towards the object, to the exclusion of all other activity. It is the mentally expressed desire for the conscious to enter into relation with the all-conscious that can perceive the object or its emanations. The expression of desire goes hand in hand with the mental orientation and reinforces it. Such expression is a powerful auto-suggestion, making the operator sensitive to the presence or emanations of the object in question, setting in motion the mental activity and giving the spirit acting in the all-conscious incredibly perceptive and selective acuity. Focus on the mental designation as expressed when formulating the desire spares the operator accidental or parasitic suggestions and random auto-suggestions, and carefully limits the action of the all-conscious to search for the designated object alone.

2. Formulation of the question
The formulation of what you want to find out needs to be expressed in the clearest manner possible. A matter of blending the information filtered carefully from the conscious mind and the designation of the object in question with the aim of the operation. The cardinal rule is clarity of expression, with no ambiguity allowed at the risk of obtaining the incorrect answer.

See the next section, "The Art of Questioning" for more on formulating the question.

3. The passive state of neutrality
This is the third stage of the operation, it is the vital condition for efficient and selective perception, as well as the surest antidote to any undesirable form of auto-suggestion. It is

a special state of mental receptivity, where you deliberately isolate the idea of what you are searching for. Nothing else exists, total inhibition of anything extraneous and distracting to the search. No matter what the object is, a material object, an abstract scientific problem, clarification of a sensitive question or a decision to be taken, this passive stage of anticipation is essential and is the upshot of concentrated attention.

Passive anticipation is necessary to establish communication with the all-conscious, which sometimes responds rapidly, sometimes slowly, and sometimes not at all. It largely depends on how you go about it, but be wary if the answer comes too quickly. Just start again calmly. You will have confirmation in repetition of the response, or by fact of likelihood (in some instances).

Immediately following the passive state of anticipation comes the active phase of the research, in fact a prolonged state of mental expectation would risk making the search sterile.

I like to think of this phase as the "mind in neutral", a state that can be easily compared to that of meditation, vacuity, rest, no wave of mentation to disturb the mind-ocean. This is allowed by the conscious mind being in such a state, so avoiding any influence while the all-conscious does the work, as usual. It is a frame of mind to be encouraged and developed because it is in this phase that the answer reaches the nerves, after a final stimulation, by:

4. Stating the question
The ground has been prepared in the previous three stages and the time is now right to ask the question once again using the formula carefully prepared in the conscious mind. Of course, this is the second occasion of asking the question, from the very start you have to know what it is you are after, and the first thing you do is to determine what you want to know and then using the four stages you filter and prepare the ground, so to speak.

One of the major principles of radiesthesia is sensitivity and this seems to be the place to mention it. You become sensitive, or rather you make yourself sensitive to something by forming the mental desire to become so. Consequently, one can become sensitive to whatever exists, in other words, you can be in osmosis esoterically. While the verbal formulation to express what the operator wants to find is of the utmost importance, it is of even greater importance to avoid exercising one's will or determination to accomplish one's aim. This is a very delicate aspect which is incompatible with radiesthesia, or natural perception, there must be no force. There is only communication with reception on the part of the operator, if one is to be in real harmony with the whole situation.

Neutrality is the secret to avoiding the influence of external suggestion and autosuggestion, simply say: "I am totally neutral, I know nothing, I am looking for...., and I would like to find...." and then continue with the questions.

Beware of: Lack of or wandering attention and forgetting any element of the four stages.

One last thing and it is very important. Don't ever perform with the pendulum in public if your intention is merely to show people how well you manipulate it. It is fine to use it in the supermarket to check if one packet of carrots is better than another producer's, but do not make an exhibition of your new-found skill. There is always someone more powerful than you. I know people who can make the pendulum swing the way they want. One can end up looking silly!

Success in a search very much depends on two things:

The strict application of the 4 Stages, and a calm state of mind during the entire search process.

Patience and perseverance are essential virtues in radiesthesia. Patience as it is required in the observation of your own physiological reactions, and assimilation to gain a firm understanding of the steps involved and your interaction. Of course, it is natural to want to press ahead at the outset of such an intriguing adventure but it is very important to study and examine. Once the operating principles have been firmly assimilated they form the grounding from which developing comprehension of the art can be anticipated and enjoyed. The key is practice, as it allows for a much better understanding of the various factors at work and their modalities of working, and confidence slowly develops.

It is definitely best initially to aim for a gradual progress, from simple to more complex matters. In other words start out with basic questions concerning:
- food,
- medication,
- everyday activities and products;
and then:
- the right-time to discuss,
- ask for a pay-rise,
- is this the right person for me,
- and so on.

-#-#-#-#-#-

Training

As previously mentioned, practice is the key to developing confidence. So, depending on how important using a pendulum is for you, it might be opportune to offer some ideas as to training and exercises.

Depending on your strengths and skills, there will be areas that could be usefully reinforced. Apart from a daily practice so as to develop confidence, the most important factor is one-pointed concentration or focus. This is no longer the virtue it once was, and there seems to be a strong chance that electromagnetic influences hamper that effort.

For those of you who practice meditation, it will not be hard to adapt to an exercise where you focus on a single thing – a small square of white paper on the wall or desk, for example. In a short period of time, you will be able to move into that limbo space, necessary for the third stage of the passive state of neutrality.

Initially, constant questioning in one's day-to-day activities provides the most immediate and convenient solution, because the answers are soon seen to be correct – or not. Such a daily practice may not be suitable, so set aside a time, specified beforehand and stick to it. Write the questions and answers down on a slip of paper so that you can check the veracity.

If you are concerned for your food intake and its quality, that is an excellent place to start. For example, you are concerned about consuming MSG or palm oil; you choose the subject. When you do your shopping, take the pendulum and ask if this brand/packet is good for you. Specify especially on each occasion if the substance is to be found in the food packet. Even if the list of ingredients does not make mention of its presence, rest assured, MSG is concealed under fifty different names!

If you suffer from a particular allergy, work on it using the pendulum. Is this soap/detergent/toothpaste good for me?

You get the idea.

Once you have acquired an ease with its use. Ask for something to be concealed in the room or garden; a key, a ring, something small on which you can focus and readily find. Once it is hidden, ask the pendulum to help you find it, by indicating the direction, you then follow the direction as the pendulum swings left/right/backwards/forwards until such time that you locate the hidden object.

Whatever happens, do not lose patience or become frustrated. Change activity if it doesn't work and come back to it later.

Having written the numbers 1, 4, 16 and 20 (or whatever you choose) on separate pieces of card-paper, place them face down on a table. Shuffle the cards and then place them in a line or a square. Take your pendulum and with a clear intention, state, "Please help me find number 4". Placing the pendulum over the first card, ask "Is this number 4?", and progress through the cards until the pendulum indicates a yes. Turn the card over and check what the number is. Expand with this sort of exercise. It is not a test, it is merely the extending of your awareness and capacity to tune in. Part of my practice was laying sixteen playing cards face down on a table, in the assembled cards was the three of hearts. I would ask to find the card. Not until I could find the three of hearts ten times out of ten was I satisfied with my ability. It is not a competition, it is a sensation. Sometimes it works, sometimes it don't!

No tension, no recrimination. Just hard work and more practice following the procedure stipulated above. I might be a poor student because it took me two years, practising about three hours a day to reach a modicum of confidence. That is not a guarantee of infallibility, but it helps! If you can, set aside a certain amount of time every day, and stick to it.

Charts, Graphs and Tables

At some stage, you are going to find a short-cut would be welcome in the questioning process, this is where charts, graphs or tables come in very handy.

Many dowsers refer to the Bovis Biometer. No doubt that the origin and the intention of this scale was brilliant. A boilermaker from Nice, André Bovis created a system to measure the vital energy of food, specifically the fruits and vegetables from his garden. He observed that their nutritional energetic value diminished over time from the moment they were picked or cut, especially leafy vegetables. He created a scale, which he named the Biometer, with a measurement spectrum ranging from 0 to 10,000, calibrated in units that he believed corresponded to angstroms (a minute measurement in the metric system, representing one tenth-billion of a meter). These later became more correctly referred to as Bovis units in circles interested in physical radiesthesia.

I would not deny the utility of and access to such precise measurement, but I find it is generally overkill for my purposes, so prefer to work on the KISS principle, I find the simplest of scales is 0-10.

That being so, the simplified scale makes for easier communication while still finding a correspondence with the Bovis scale. He found that the minimum nutritional value for a food was to be found at 6,500 BU; likewise, 6.5/10 is often found to be the cutoff mark between 'good' and 'bad'.

The chart below was made for me graciously by Phil Claffey and incorporates a number of factors that I use in analysis of the diverse aspects of the things I work on.

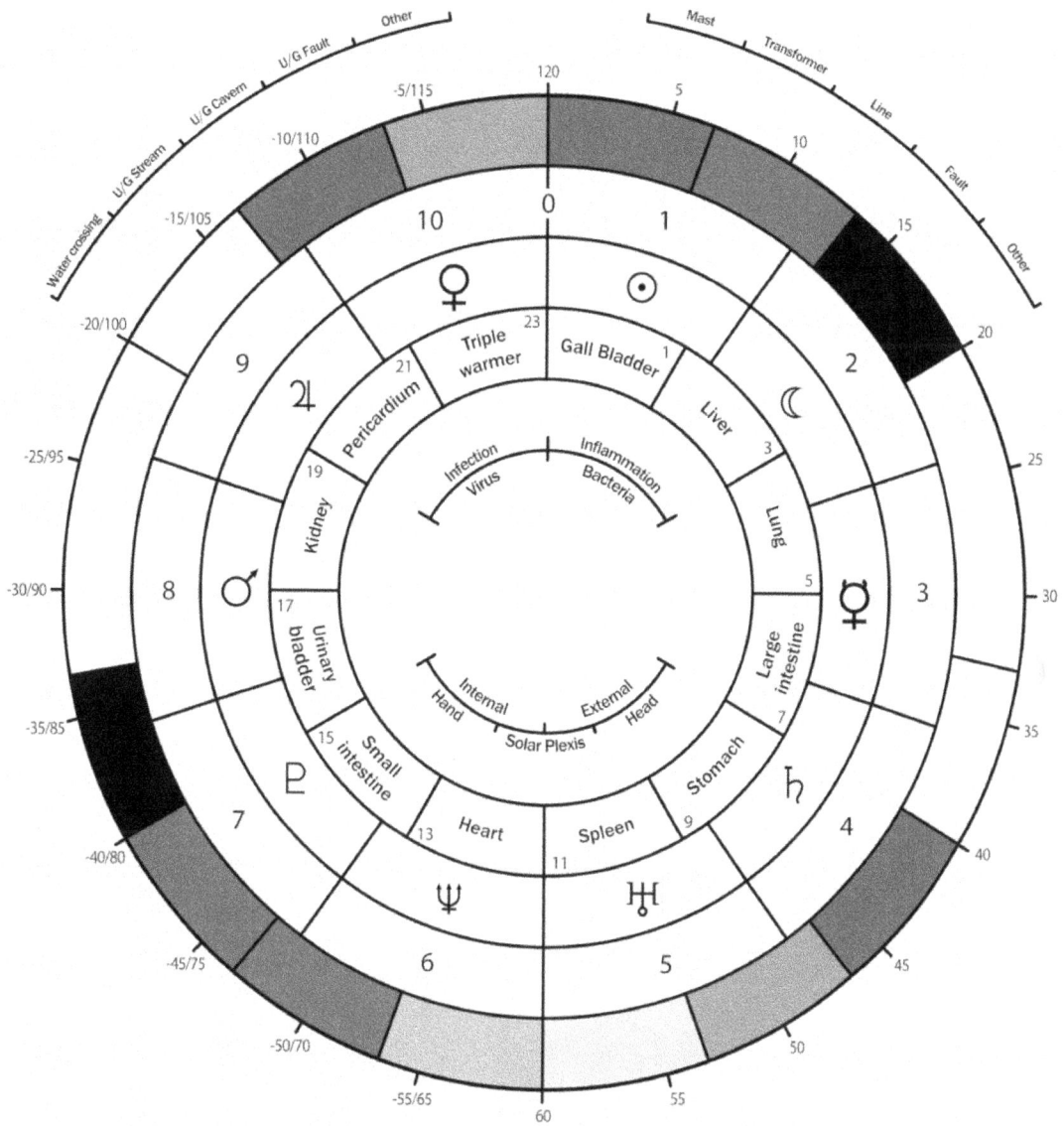

White, Indigo, Blue, Purple, Green and Black at the top.
Black, Neg.Green, Red, Orange, Yellow, Blue, Indigo, White at the bottom.
And again, another chart with more criteria included.

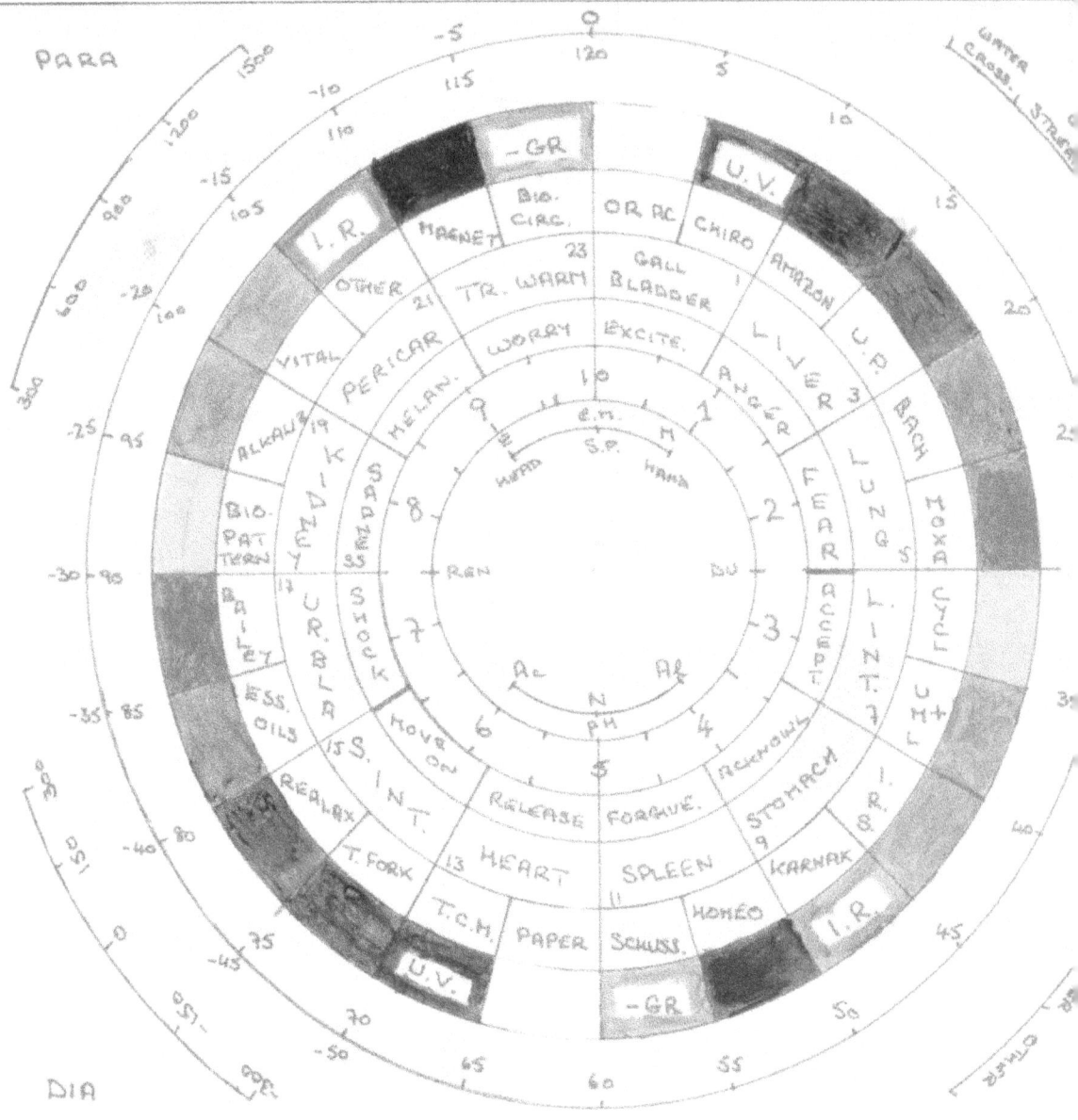

These are very personal instruments and are best made-to-measure as a function of what you do, your applications and knowledge base.

The essential in matters of charts being, I believe, a shortcut to questioning. There is no point finding out all sorts of arcane data if you cannot apply it.

As you are beginning to appreciate, my whole approach to radiesthesia is practical.

APPLICATIONS

What do you do when confronted with a dilemma? You have worked through the issue in as logical a manner as possible, but the answer is still obscured. This is where the pendulum comes into its own; and confidence is the key. It is a little known fact that medical doctors (about three thousand of them) in France during the 1950s were using the pendulum to seek confirmation of their diagnosis and correct prescription, until the national medical authorities passed the order down to cease the habit or face the risk of being debarred from the profession. History does not recount whether this was the work of the increasingly powerful pharmaceutical lobby, or an aversion to non-allopathic methods. It is, however, indicative of the confidence given by an entire strata of society to the practical use found in this eminently simple method.

It is probably best to use your expertise and/or interests as the foundation for your future development in the application you opt for, but realize that there is endless potential in the use of a pendulum.

Some of the more common applications include:

1.   The search for water, oil and mineral deposits. This is the well-known traditional field of dowsing.
2.   Archaeological exploration. A more limited field but of considerable and increasing importance for historical research and the recovery of vanished prehistoric remains (remember to include private collections as an option in the search for a relic).
3.   Architectural uses, such as site dowsing, in which must be included detection of harmful earth rays and detection of cavities, pipes and drains etc. For the Chinese with their *feng shui*, no dwelling should be built until the site has been properly dowsed. The actual building materials are also important, as are the substances used in furniture; steel, for example, dulls the brain - it is a mineral hypnotic.
4.   The location of criminals, missing persons, dead bodies, and lost or buried property and money.
5.   Agricultural and horticultural uses. In such things as the determination of optimum soil conditions, seed fertility and germination, plant health, presence of mycorrhiza, and of good husbandry in general including the value of all additives both organic and inorganic. Determination of quality, aliveness and wholesomeness in all foods whether natural, manufactured, processed, or artificial and synthetic.
6.   Character Analysis or Personality assessment, along the lines of Galen, Paracelsus and Dr. Oscar Brunler. It has manifold uses, educationally and industrially, in estimation of talents, aptitudes, personality problems and mental potential, etc.
7.   Medical and Veterinary application. Be careful here, as it is illegal to practise medicine without a licence and the industries involved do not appreciate us 'labourers' meddling in

their backyard, even if the proof is there that alternatives produce concrete results. You don't need to be measured out for concrete boots!

8.   Homoeopathy. Radiesthesia has been used in the practice of homoeopathy for many years now with great success.

9.   Environmental issues. Research into pollution and contamination can be much facilitated thanks to dowsing, especially when searching for remedies. Thanks to radiesthesia I came upon the orgone accumulator to alleviate nuclear irradiation, and the FDV to offset electromagnetic issues caused by mobile telephony.

10. Research. Whether of a historic or societal nature.

Confidentiality and respect of belief

It goes without saying that you will become privy to confidential information in your work. Rest assured, no one will believe you if you report a coven of witches and warlocks involved in nasty behavior down the way, so it can sometimes become quite complex as to how one should react. Probably the best solution is to ask with the pendulum, if the situation becomes too much for your conscience.

On a simpler note, keeping trust and respecting privacy is essential, and is to be expected of a radiesthesist. Much like a member of the professions (medical, legal, accounting, translation, etc.), you must keep mum about what you learn; not because you are not expected to shout from the rooftops about what you have discovered concerning the neighbour's private life, but because intuition works in subtle ways and can only be developed in compassion and love. If you betray that privileged understanding, it is you who will suffer.

Obviously, there are cases when you are totally out of your depth and have no experience of what is confronting you. These are times to learn, even if you do not understand what is happening, let alone why you, and why now. Play the pendulum more than ever. That is how one refines the questioning and additional insights into the esoteric are achieved; in the knowledge that every context is unique and never to be encountered again.

Seeking help from trusted external sources, so long as confidentiality is maintained, is to be encouraged and often the only way to progress.

There is a useful dictum to have constantly in mind: 'There is always some one or thing stronger than yourself.' This may sound presumptuous and sounds as if you think you have some power or other. You do, but not in the conventional sense. You are the guardian of a very personal and private connection with the all-conscious; it is to be treasured and nourished. That is the ultimate guide.

PROTECTION

This brings us to the question of protection. As you will discover in the chapter on bio-magnetism, it is my understanding that we humans are magnetic creatures, subject to forces that we have learned to refer to as 'electric', 'magnetic', and so on. While it is most useful to have a peg on which to hang our beliefs and explain them in our scientific jargon, so that we are reassured and secure in our knowledge, I don't adhere to such a credo. The pendulum will answer if you ask questions along these lines, unfortunately the answers change and are cause for constant confusion to our received scientific education. At least that is my experience, and so be it. What is important, I think, is to find solutions, and preferably natural. Very often these solutions assume a geometric form.

Enter the Atlantis symbol:

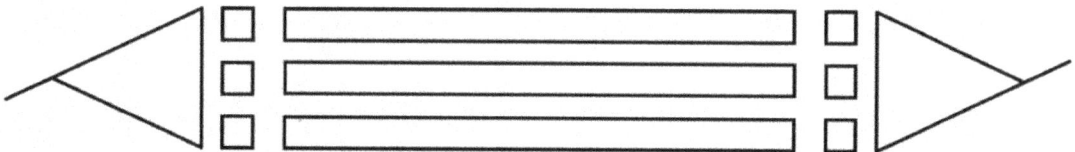

Which needs to be tweaked to become effective in its role as enhancer of the magnetic field, and therefore, as protection.
In its ring form:

Some background on the ring:

The Atlantis symbol is perhaps the world's oldest "magical" **geometric form to have reached our modern age** and a very powerful one too. Powerful in what it can achieve as regards protection of a quite specific kind, but it is only effective if the design specifications are met, otherwise it serves little or no purpose. Great attention must be paid to the geometry and proportion.

We know nothing about its origins, so theories and rumour abound!

Using radiesthetic procedures, however, it can be established that the combination of the geometric patterns and forms making up the Atlantis symbol harmonize telluric and cosmic energies, but offer no shield against cosmic forces emanating from the sun and universe, nor EMF or radiation. Basically, that means the effect can be used as protection from a large number of magnetic forces, both known and unknown influences, defending the wearer from energy originating in the earth. That is why it is especially useful as a defence against 'psychic attack', 'black magic', harmful intention of both human and mane, providing protection of a psychic and mental nature. It functions on a permanent basis, 24 hours a day with a sphere of influence measuring approximately 2.5 metres. The symbol generates a magnetic energy and boosts the vital force of the wearer.

The overall pattern produces a magnetic field, projecting a negative magnetic polarity in the axis of the centreline for a distance of 60 cm or so in both directions, depending on the energetic makeup of the immediate environment (including the wearer's magnetic field). The geometric forms are of great importance in creating the polarized field, with each component playing a role. The proportion of the pattern in relation to the overall field is of great importance, to the extent that if it is not respected, the ring serves no purpose at all. The central point of the three parallel lines is neutral with the outside edge of the ring producing a positive magnetic polarity. The groove connecting the two holes on the inside of the ring ensures the complete loop of the magnetic circuit. The ring does not generate vital force but conveys it on the magnetic phase of the field created by the ring.

In the domain of black magic, where hexes or curses are common coinage, this pattern is of great value. I wouldn't know about the "curse of Tutankhamen" but apparently Howard Carter owned a sandstone ring with the symbol engraved on it. That was the start of my research into the capacities of the ring and it proved to be totally efficient in fending off attacks from what I can only call, for lack of adequate understanding and vocabulary, determined demonic beings. Both on other people and myself.

Earlier mention was made of the work of André de Belizal, Léon Chaumery and P.A.Morel. They experimented with some powerful energies, including a very low frequency, going by the patented name of Negative Green, that de Belizal says took the life of Chaumery by dehydration – one of the effects of this specific energy form as seen with mummification in a pyramid. The great-grandfather of de Belizal's wife, the Marquis d'Agrain, an Egyptologist on Napoléon's expedition to Egypt in 1804, had "found" a sandstone ring with the Atlantis symbol engraved on it. This ring was permanently in a drawer of de Belizal's desk during the period of their experimentation in the 1930s and 40s, that spared André de Belizal the dehydration problem that took the life of his colleague.

It is highly unlikely that this was the same ring as Carter's, given that de Belizal's ring had been a family heirloom since the early nineteenth century and is still with the family.

Carter was an active trader in Egyptian artifacts during the time he spent in Egypt, which was before de Belizal was experimenting. The rumour, and it was nothing more than that, of Tutankhamen's curse which supposedly caused the untimely deaths of members of Carter's excavation, was created by sensation-hungry journalists. The only person who died shortly after the uncovering of the tomb was Lord Carnavon, the expedition's sponsor. His death was closely related to his presence in Egypt where he had gone on doctor's advice to stay in a dry climate following a car-racing accident in the UK, when he had seriously damaged his lungs. He died following a blood infection after he had nicked a mosquito bite whilst shaving. Not quite the type of news to reach the headlines!

There are doubtless other forms of protection – amulets, crystals, herbs, mantras, rituals, and so on. Ask with the pendulum what is the best for you and those you work with.

END

About the author

The basis of this text is founded essentially on a combination of practical experience, the study of theory from selected books, substantial personal practice and experimentation in a variety of methods and traditions. It all started in India in the early seventies when I studied Sanskrit so as to read the ancient texts in the original rather than relying on translation. In that phase, I had the opportunity to study with Swami Pranav Tirtha, a *dashnami sannyasin*, who initiated me into the Vedanta philosophy. Whilst with him I read, studied and assimilated the orthodox teachings of the Upanisads, the Brahma Sutra, Gita and multiple metaphysical and sundry texts of Hindu literature. I was ordained as a monk, with the name of Swami Chidananda Tirtha in May 1973. This period also furnished the occasion to study medicine with Dr Himatlal Trivedi, an Ayurvedic practitioner from Palitana, whom I accompanied in India and Africa in his practice amongst English and Gujerati-speaking patients. That involved study of the Hindu medical classics (Caraka Samhita, Sushruta Samhita, Ashtanga Hrdaya), with considerable practical experimentation of fasting and dietary regime on myself. Observation with Himatlal's guidance and explanation gave me a reasonable understanding of this medical art form.

Whilst living in France, I had the opportunity to study for a year with a French acupuncturist, who was persuaded to come out of retirement to teach Traditional Chinese Medicine (TCM) again. That grounding was then followed by many years of studying the Chinese classics: the Lingshu, the Huainanzi, the Suwen, along with in-depth reading of Soulié de Morant, Claude Larre and Elisabeth Rochat de la Vallée, in addition to extensive practice of moxibustion and acupressure. Whilst living in Chiang Mai, the opportunity arose to learn and practice a form of bio-energetics which involved a lot of practical moxibustion. The outcome of my studies of TCM affords a certain ease with this very complete approach to the human condition.

A vast amount of research, reading and experimentation with a very broad spectrum of subjects, traditions and cultures, combined with travelling and living among natives of other lands, along with my professional activity as a technical translator, specialized in nuclear and telecom technologies, for some twenty-five years in France have hopefully been turned to good advantage.

Study, practice and research into magnetism, laying on of hands, geomancy, radiesthesia and radionics add to my wholistic comprehension of life. A practice of organic farming combined with animal husbandry, special care for water and its supply have also led to my current understanding. I work more with the intention of clarifying what I think seems to be happening in the dimensions we evolve in, rather than any kind of dogmatic laying down of law.

Whilst I sincerely believe my opinions to be correct because corroborated by the pendulum and experience, it would be a substantial error to think this is the last word because it concerns uniquely what falls within my own sensory (all three hundred and sixty) parameters. Too much is changing too fast for our perceptual ability to stay abreast of events, even were we able to comprehend and adapt. Perhaps it is not for us humans to determine how the ordered immensity of Nature works; that would be most presumptuous and dishonest, but it does seem to be worth trying to establish a mode of operation that might serve as a guide or possible reference fitting into our journey through this phenomenal existence.

I stand on the shoulders, hopefully in rectitude and fidelity to the thrust of the original argument of many researchers, practitioners and remarkable people in aligning these words on paper. The words will, as always, be symbols of the generosity of Nature as she carefully keeps everything in its structured place, although the human component must be the most unruly, hence the hard lessons we have to learn if we aspire to some other form of existence than the purely material.

www.ingramcontent.com/pod-product-compliance
Lightning Source LLC
Chambersburg PA
CBHW080631030426
42336CB00018B/3159